SCIENTIFIC AMERICAN EDUCATIONAL PUBLISHING

SCIENTIFIC AMERICAN INVESTIGATES SPIDERS

BLACK WIDOWS

BY ROXANNE ALDOLPHO

Published in 2025 by The Rosen Publishing Group
in association with Scientific American Educational Publishing
2544 Clinton Street, Buffalo NY 14224

Library of Congress Cataloging-in-Publication Data
Names: Aldolpho, Roxanne, author.
Title: Black widows / Roxanne Aldolpho.
Description: [Buffalo, NY] : Scientific American Educational Publishing, an imprint of Rosen Publishing, [2025] | Series: Scientific American's library of spiders | Includes bibliographical references and index.
Identifiers: LCCN 2023059114 | ISBN 9781725351189 (library binding) | ISBN 9781725351172 (paperback) | ISBN 9781725351196 (ebook)
Subjects: LCSH: Black widow spider–Juvenile literature. | Spiders–Juvenile literature.
Classification: LCC QL458.42.T54 A43 2025 | DDC 595.4/4–dc23/eng/20240123
LC record available at https://lccn.loc.gov/2023059114\

Portions of this work were originally authored by Jill Keppeler and published as *Black Widows.* All new material in this edition is authored by Roxanne Aldolpho.

Designer: Andrea Davison-Bartolotta
Editor: Brianna Propis

Photo credits: Cover phonecat/Shutterstock.com; p. 5 Sari ONeal/Shutterstock.com; p. 7 (map) trgrowth/Shutterstock.com; p. 7 (cartoon spiders) Webspark/Shutterstock.com; p. 7 (bottom) Gayle Marien/Shutterstock.com; p. 9 Jay Ondreicka/Shutterstock.com; p. 11 Malpolon/Shutterstock.com; p. 13 Jeff W. Jarrett/Shutterstock.com; p. 15 Sari ONeal/Shutterstock.com; p. 17 Pong Wira/Shutterstock.com; p. 19 (main) Nick626/Shutterstock.com; p. 19 (web) Chinch/Shutterstock.com; p. 21 Protasov AN/Shutterstock.com.

Printed in the United States of America

CPSIA compliance information: Batch #CSSA25. For Further Information contact Rosen Publishing at 1-800-237-9932.

CONTENTS

Words in the glossary appear in **bold** type the first time they are used in the text.

WHO'S AFRAID OF THE BLACK WIDOW?

The black widow spider doesn't look like the scariest spider in the world. It doesn't bother humans, and it's only the size of two of your fingertips! So why is the black widow spider considered one of the scariest spiders in the world? One reason: its **venom**.

The venom of a black widow is deadly. However, these small, shy spiders don't want to hurt people. They would much rather stay away! But it's still important to be careful if you're ever near one.

Black widow spiders are arachnids. Arachnids are animals that have an **exoskeleton**, a **segmented** body, and eight jointed legs.

FUN FACT

BLACK WIDOW SPIDER VENOM IS 15 TIMES STRONGER THAN RATTLESNAKE VENOM!

NIGHT CRAWLER

There are a variety of different species, or types, of black widow spiders. They live many places on Earth, including throughout the United States. They especially like areas where the weather doesn't get too hot or cold, such as in the southern United States.

Black widows often make their webs in dark, covered places that are close to the ground, such as under logs or stones or in pipes. These spiders are very shy. They're nocturnal, or most active at night, and typically spend time by themselves.

Where Black Widows Live

The average life span of black widows in the wild is one to three years.

FUN FACT

BLACK WIDOWS TEND TO BUILD THEIR WEBS IN SMALL HOLES AND CRACKS OUTDOORS. KEEP AN EYE OUT FOR THEM IN YOUR BASEMENT, ALTHOUGH IT'S UNCOMMON TO FIND THEM INDOORS.

IDENTIFYING THE WIDOW

Female black widow spiders can be **identified** by their bodies and markings. Their bodies are shiny and black. They usually have a bright red or orange hourglass-shaped marking on the bottom of their **abdomen**. Males are lighter in color and have reddish spots on their back. Sometimes they have stripes.

A female black widow spider can be about 1 inch (2.5 cm) long. Male black widow spiders are much smaller, about one-fourth to one-half the size of the females. However, people are more likely to see a female black widow than a male.

Can you spot the hourglass shape on this female black widow spider? It looks a bit like two triangles together.

FUN FACT

MALE BLACK WIDOW SPIDERS DO NOT BITE, BUT FEMALES DO. THEY BITE OTHER BLACK WIDOWS OR PREDATORS TO GUARD THEIR EGGS.

HUNGRY FOR MORE

Have you ever wondered where black widow spiders got their name? It comes from an interesting **behavior** they will sometimes show! Although black widows usually live alone, they do come together to mate, or make babies, in the late spring.

After the spiders mate, the female black widow sometimes kills and eats the male spider! This doesn't always happen. But it happens often enough for the black widow spider to get its name and **reputation**. "Widow" is the name for a woman whose husband or wife has died.

Female black widows have been observed eating their mates in **captivity**, but rarely in the wild.

FUN FACT

SOME MALE BLACK WIDOWS FLING THEMSELVES OFF THEIR MATES RIGHT AFTER MATING TO KEEP FROM BEING EATEN!

WHITE WEBBING

The webs spun by black widows are large, white, and messy! These webs may look messier than the webs that many other spiders make. Sometimes they even look tangled or torn. However, the webs are actually very strong and well planned.

Female spiders create round silk egg cases in their webs. Each case is about 0.5 inch (1.3 cm) wide and can hold 250 to 750 tiny spider eggs. The female may lay eggs several times over one summer, bringing lots of baby black widows into the world!

This black widow has an egg case holding its unborn babies in its web. An egg case feels a bit like paper.

FUN FACT

BLACK WIDOWS SPEND A LOT OF TIME ON THEIR WEB AWAY FROM OTHER CREATURES.

BABY SPIDERS

Baby black widow spiders, called spiderlings, come out of their eggs after around 14 to 30 days. When black widow spiderlings are first born, they're orange and white. Both males and females look more like adult male black widow spiders. Baby spiders aren't harmful to humans.

The young spiders leave the web within a few days, spinning bits of silk that catch the wind and carry them to new areas. This is called ballooning. Once grown, black widow spiders can live for up to three years, but around one year is more common. Females tend to live longer than males.

FUN FACT

BLACK WIDOW SPIDERLINGS ARE CANNIBALISTIC, MEANING THEY WILL EAT THEIR SIBLINGS FOR FOOD. THE SURVIVORS LEAVE THEIR MOTHER'S WEB AFTER A FEW DAYS.

These spiderlings climb on the web while their mother guards them. The baby females will turn black as they get older.

THE MENU

When insects get stuck in a black widow's web, they become food for the spider. Black widows like to dine on flies, mosquitoes, grasshoppers, beetles, and other small creatures.

The spiders have an interesting way of eating. Once a black widow has caught an insect in its web, its **fangs** make holes in the insect's body. Then it shoots a special matter into the insect's body to turn the insides into liquid—and it sucks that liquid into its mouth. It's like insect soup. What a yucky menu item!

An insect has been trapped in a black widow's web and will soon be eaten! Adult black widows only eat every few weeks or even once a month.

FUN FACT

BLACK WIDOW SPIDERS HAVE SHORT, DARK HAIRS COVERING THEIR BACK LEGS. THESE HAIRS HELP THEM TO COVER THE INSECTS THEY CATCH WITH SILK.

ALL BITE, NO BARK

Black widows are feared because of their venomous bites. It's important to tell an adult as soon as possible if you think a black widow spider might have bitten you. Black widow venom can kill people!

The good news is that most people will get better after a bite if they're treated quickly by a doctor. Black widow spiders don't always use venom when they bite—but it's still a good idea to be careful.

FUN FACT

A BLACK WIDOW'S BITE WILL LEAVE BEHIND SWELLING AND REDNESS, AS WELL AS TWO SMALL FANG MARKS THAT LOOK LIKE TINY RED DOTS.

A black widow's bite can cause sweating, headaches, and even muscle pain.

Be Careful!

HERE ARE SOME IMPORTANT TIPS FOR SAFETY AROUND BLACK WIDOWS—AND HOW TO KEEP FROM GETTING BIT!

- Shake out blankets or clothes stored in dark, closed-up places before using them.
- Look into dark places where spiders might have webs before reaching into them.
- Don't poke at spiders! They'd rather you stay away.
- If a spider bites you, tell an adult right away. If possible, an adult should to try to catch the spider and take it to the doctor with you.
- Look for the spider's red hourglass marking.
- Stay away from piles of rocks or wood. Be careful in areas where the spiders might like to spin webs. Wear gloves if you are working in these areas.
- Shake out your shoes before putting them on if they've been in a garage or someplace dark.

MORE HELPFUL THAN HARMFUL

Black widow spiders can seem very scary. With their strong venom and black bodies, they may cause the hairs on the back of your neck to stand up! But remember, these shy spiders really don't want to bite people. They'd rather use their venom on tasty insects.

Black widows play an important role in their **ecosystem**. They control insect populations and eat pests that destroy people's crops and gardens. People and black widows can live side by side. We just have to make sure not to upset their peace!

The black widow spider may be useful in **biotechnology**. Its venom could be used to create new **insecticides**, or even medicines!

FUN FACT

BLACK WIDOW SPIDER BITES ARE MOST DEADLY TO CHILDREN, OLDER PEOPLE, AND PEOPLE WHO ARE ALREADY SICK.

GLOSSARY

abdomen: The part of a spider's body that contains the stomach.

behavior: The way an animal acts.

biotechnology: Technology that uses living organisms or their creations to make new products.

captivity: The state of being caged.

ecosystem: All the living things in an area.

exoskeleton: The hard outer covering of some kinds of animals.

fang: A hard, sharp-pointed body part a spider uses to put venom into its prey.

identify: To find out the name or features of something.

insecticide: A substance used for killing insects.

reputation: The views that are held about something or someone.

segmented: Divided into different parts.

venom: Something an animal makes in its body that can harm other animals.

FOR MORE INFORMATION

Books

Anthony, William. *Black Widow Spider.* New York, NY: Enslow Publishing, 2022.

Murray, Julie. *Animals with Venom: Black Widow Spider.* Mendota Heights, MN: North Star Editions, 2020.

Websites

Hey! A Black Widow Spider Bit Me!
kidshealth.org/en/kids/black-widow.html
You can never be too careful—read about what to do if you're ever bitten by a black widow.

National Geographic Black Widow Spiders
www.nationalgeographic.com/animals/invertebrates/facts/black-widow-spiders
Learn more about the lifestyles of the shy but deadly black widows.

Publisher's note to educators and parents: Our editors have carefully reviewed these websites to ensure that they are suitable for students. Many websites change frequently, however, and we cannot guarantee that a site's future contents will continue to meet our high standards of quality and educational value. Be advised that students should be closely supervised whenever they access the internet.

INDEX